Oceans Of The World In Color

Speedy Publishing LLC
40 E. Main St. #1156
Newark, DE 19711

www.speedypublishing.com

Copyright 2014
9781635011128
First Printed October 28, 2014

All Rights reserved. No part of this book may be reproduced or used in any way or form or by any means whether electronic or mechanical, this means that you cannot record or photocopy any material ideas or tips that are provided in this book.

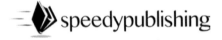

Ocean Facts...

Around 70% of the Earth's surface is covered by oceans.

Ocean Facts...

While there are hundreds of thousands of known marine life forms, there are many that are yet to be discovered, some scientists suggest that there could actually be millions of marine life forms out there.

Ocean Facts...

Ocean tides are caused by the Earth rotating while the Moon and Sun's gravitational pull acts on ocean water.

Ocean Facts...

Located to the east of the Mariana Islands in the western Pacific Ocean, the Mariana Trench is the deepest known area of Earth's oceans. It has a deepest point of around 11000 metres (36000 feet).

Ocean Facts...

Oceans are frequently used as a means of transport with various companies shipping their products across oceans from one port to another.

Ocean Facts...

The deep sea is the largest museum on Earth: There are more artifacts and remnants of history in the ocean than in all of the world's museums, combined.

Ocean Facts...

We have only explored less than 5 percent of the Earth's oceans. In fact, we have better maps of Mars than we do of the ocean floor (even the submerged half of the United States).

Ocean Facts...

At that depth, the temperature is always just above freezing, the pressure is more than 1000 times what it is on the surface, and many bottom-dwelling fish and invertebrates call it home!

Ocean Facts...

If all the ice in glaciers and ice sheets melted, the sea level would rise by about 80 meters (262 ft), about the height of a 26-story building.

Ocean Facts...

If sea level should rise by 3 meters (10 feet), many of the World's coastal cities, like Venice, London, New Orleans, and New York, would be under water.

Ocean Facts...

The largest ocean on Earth is the Pacific Ocean, it covers around 30% of the Earth's surface. The Pacific Ocean's name has an original meaning of 'peaceful sea'.

Ocean Facts...

The Pacific Ocean is surrounded by the Pacific Ring of Fire, a large number of active volcanoes.

Ocean Facts...

The second largest ocean on Earth is the Atlantic Ocean, it covers over 21% of the Earth's surface.

Ocean Facts...

Amelia Earhart became the first female to fly solo across the Atlantic Ocean in 1932.

Ocean Facts...

The third largest ocean on Earth is the Indian Ocean, it covers around 14% of the Earth's surface.

Printed in Poland
by Amazon Fulfillment
Poland Sp. z o.o., Wrocław